ESSENE COMMUNIONS WITH THE INFINITE

LIGHTNING FLASHES IN THE GATHERING STORM

by
EDMOND BORDEAUX SZEKELY

*"Everything lives,
and everything is holy."*
—*Endre Ady*

MCMLXXIX
INTERNATIONAL BIOGENIC SOCIETY

SOME BOOKS BY EDMOND BORDEAUX SZEKELY

THE ESSENE WAY—BIOGENIC LIVING
THE ESSENE GOSPEL OF PEACE, BOOK ONE
BOOK TWO, THE UNKNOWN BOOKS OF THE ESSENES
BOOK THREE, LOST SCROLLS OF THE ESSENE BROTHERHOOD
BOOK FOUR, THE TEACHINGS OF THE ELECT
THE DISCOVERY OF THE ESSENE GOSPEL: The Essenes & the Vatican
SEARCH FOR THE AGELESS, in Three Volumes
THE ESSENE BOOK OF CREATION
THE ESSENE JESUS
THE ESSENE BOOK OF ASHA
THE ZEND AVESTA OF ZARATHUSTRA
ARCHEOSOPHY, A NEW SCIENCE
THE ESSENE ORIGINS OF CHRISTIANITY
TEACHINGS OF THE ESSENES FROM ENOCH TO THE DEAD SEA SCROLLS
THE ESSENES, BY JOSEPHUS AND HIS CONTEMPORARIES
THE ESSENE TEACHINGS OF ZARATHUSTRA
THE ESSENE SCIENCE OF LIFE
THE ESSENE CODE OF LIFE
THE ESSENE SCIENCE OF FASTING AND THE ART OF SOBRIETY
ESSENE COMMUNIONS WITH THE INFINITE
THE FIRST ESSENE
THE BIOGENIC REVOLUTION
THE ORIGIN OF LIFE
THE COSMOTHERAPY OF THE ESSENES
THE LIVING BUDDHA
SIDDARTHA, THE PILGRIM
PILGRIM OF THE HIMALAYAS
MAN IN THE COSMIC OCEAN
TOWARD THE CONQUEST OF THE INNER COSMOS
FATHER, GIVE US ANOTHER CHANCE
THE ECOLOGICAL HEALTH GARDEN, THE BOOK OF SURVIVAL
THE TENDER TOUCH: BIOGENIC FULFILLMENT
THE DIALECTICAL METHOD OF THINKING
THE EVOLUTION OF HUMAN THOUGHT
THE SOUL OF ANCIENT MEXICO
ANCIENT AMERICA: PARADISE LOST
MESSENGERS FROM ANCIENT CIVILIZATIONS
SEXUAL HARMONY: THE NEW EUGENICS
LUDWIG VAN BEETHOVEN, PROMETHEUS OF THE MODERN WORLD
THE FIERY CHARIOTS
CREATIVE WORK: KARMA YOGA
THE ART OF STUDY: THE SORBONNE METHOD
COSMOS, MAN AND SOCIETY
THE BOOK OF LIVING FOODS
SCIENTIFIC VEGETARIANISM

Cover: *Jesus*, by Leonardo da Vinci. Brera Gallery, Milan.

Book Design by Golondrina Graphics

I dedicate this book to the author, Edmond Bordeaux Szekely, whose untiring efforts during the last half a century have given us the priceless knowledge of the Essene teachings.

–Norma Nilsson Bordeaux Szekely

How could the love between Thee
and me sever?
As the leaf of the lotus abides on
the water; so Thou art my Lord,
and I am Thy servant.
As the night-bird gazes all night at
the moon: so Thou art my Lord
and I am Thy servant.
From the beginning until the ending
of time, there is love between
Thee and me; and how shall such
love be extinguished?
As the river enters into the ocean,
so my heart touches Thee.

–Kabir

Preface

This precious jewel of a book is the fortunate result of my own "archeological excavations." Going through the papers of Dr. Szekely a few days after his death, I found a manuscript in his own hand, with instructions to me, about Essene traditions from original Zarathustrian sources, concerning certain regenerative communions which allowed them to commune with and return to their original ancestral harmony with the Eternal Streams of Life, Sound, and Light. The awe I felt as I read it came not only from the spiritual power of the manuscript, but also from intense gratitude that I am the only person on this planet who can read his handwriting! I copied it word for word, experiencing in the process both heights of joy and sorrow: joy, because this book represents the very essence of the hidden Essene teachings, delving into the esoteric mystery of the communions with the Lifestream, the Soundstream, and the Lightstream; sorrow, because, after all, it is only a month since he left us in his physical form, and I love him, and I am only human.

Nevertheless, I and all who treasure his books are luckier than most who lose a loved one. We have his works, the magnificent outpouring of one of the greatest minds of this century, and perhaps of all time. His books will always be with us, and in his books, he lives.

And we must do our part to *practice* what is in this book, not only read and admire its beauty. He could not have given us a more cherished parting gift than the knowledge and esoteric wisdom hidden in these Communions with the Holy Streams of Life, Sound, and Light. Behind their seeming simplicity lies an unknown realm of Truth and Reality—that Home for which every Essene Wayfarer longs. We have a sacred duty to follow the Path which he so faithfully followed all of his long, long life.

—NORMA NILSSON BORDEAUX SZEKELY

Orosi, Costa Rica, September, 1979.

CONTENTS

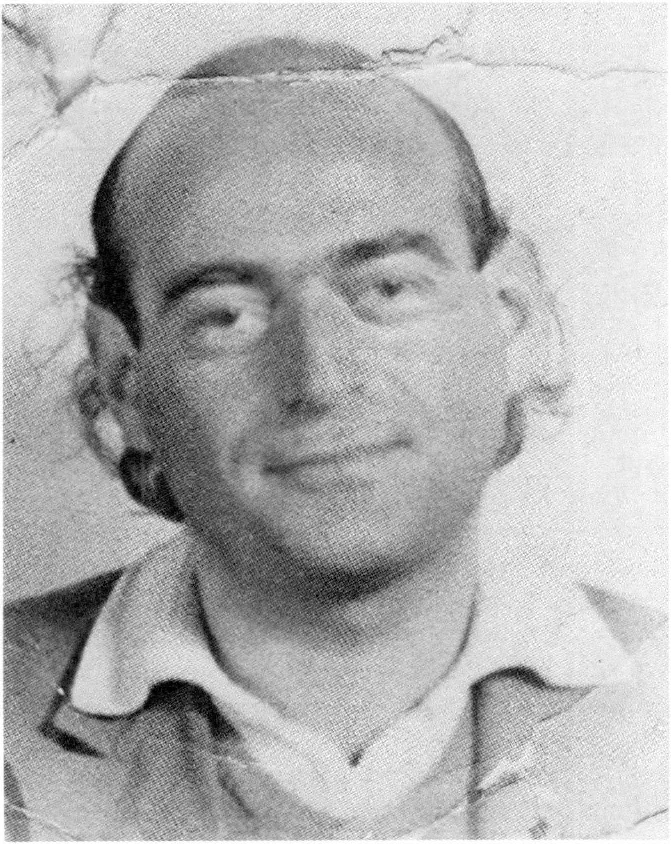

Edmond Bordeaux Szekely *(circa 1945)*

PROLOGUE
At the Grand Central Station

From the beginning of the twentieth century, a slow, gradual, quasi-imperceptible process began to evolve, creating a greater and greater impact, in fact, the greatest in history, on the life of Western man. Everything connected with our daily lives is becoming more and more complex, more and more differentiated and more and more complicated, affecting our physical, mental, social and economic environments. By the inexorable and inevitable trend of the progress of technological civilization, Western man has created such enormous complexities of life that his nervous system has started to disintegrate under the weight and stress of these complexities. The omnipresence of stress is a specific twentieth century phenomenon. If we analyze the origins of this stress with its concomitants of nervous tension, insecurity and gradual inability to cope with problems, we will find the following factors: greater and greater complexities in time, in place, in human activities, in the speed of these activities, in mental activities, in emotional activities, and in bodily degenerative processes. For instance, in previous civilizations, people usually lived in the same area their whole lives, without the necessity of constant travel in the performance of their daily functions. In previous civilizations, the activities of people were few and simple and did not extend to a great number of diversified activities and problems, in contrast to Western man, whose daily timetable is parceled into countless heterogeneous fractions of activities. In previous civilizations, the speed of all these activities in time and space was quiet, slow, and not so exacting. In previous civilizations, the mental functions of man were also very simple and concerned only the few basic necessities and aspirations of life, without the need of continuous calculations, preoccupations, decisions and solutions of innumerable problems.

In previous civilizations, the emotional life of man was rather uncomplicated, contrasting severely with the great number of increased emotional reactions of a more and more complex society and human environment of today. Finally, the last important factor in the origins of stress, the bodily degenerative processes, are accelerated by a sedentary way of living, punctuated by abnormal spurts of the wrong kind of exercise, and aggravated by a continuous source of synthetics from the air we breathe and the foods we eat.

It seems that stress, with its concomitant factors of nervous tension and insecurity, has become the greatest problem facing Western civilization in the middle of the twentieth century. We are not concerned with that minority who need professional treatment, nor with the other minority of those who enjoy perfect mental health, but with the vast majority who, in between the two minority groups, are not so mentally ill as to need professional care, nor yet are completely healthy mentally. I speak of the great majority under continuous or periodical stress, with nervous tension and feelings of insecurity. This majority does not yet need professional help, only a simple, preventive program which they could practice themselves.

It is evident that we cannot turn back the clock, nor the trend of progress, evolution and history. We cannot go back to the natural simplicity of life in previous civilizations, and we cannot eliminate the continuously increasing complexities imposed by the inexorable forces of constant expansion of human knowledge and activities. No unrealistic, Utopian approach can give us the solution to the problem of stress. The only realistic approach which remains for us is to realize the origins of stress, which are inseparably and inevitably connected with our twentieth century Western civilization. We must realize and accept the fact that stress is as inevitable as the factors which are causing it. But if we cannot eliminate stress completely, at least we can alleviate it and decrease it by excluding the stress-causing factors of our century, from time to time, trying to escape from them periodically and returning now and then to the non-stress-forming practices of previous civilizations. It is possible, even in the twentieth century, to reconstruct and enact all the beautiful practices of ancient civilizations, achieving for us at least a temporary immunity to the stress-causing factors in our hectic lives. And there is tremendous psychosomatic value in recharging our batteries from the great sources of harmony and knowledge of previous ages. We want to emphasize that while it is a mistake *not* to live in the twentieth century, it is an even greater mistake to live *only* in the twentieth century.

Every so often, we should try to take a mental vacation from the twentieth century, a respite which may eventually become our means of liberation from all stress-causing factors.

From time to time we should exclude the multiple stimuli of our stressful environment: the visual ones, such as television, billboards, headlines, neon signs; and the acoustic ones, such as television, telephones, radios, inter-coms, traffic, machines, all the sounds of daily turmoil which constantly fill our minds with haphazard and irrelevant contents. Through this periodic holiday from the present, we will free ourselves from the continuous slavery of permanent external stimuli, and finally become masters of our own minds—in one word, we become ourselves. We regain our own personality, which, in the words of Goethe, is the greatest gift of the gods to humans. Only when the ephemeral noise of the Grand Central Station around us is transcended can we accomplish this ultimate freedom of the mind.

And once we have successfully created a vacuum in our minds, we will be able to fill them with contents of our own choosing, according to our inner dispositions. We can utilize all the finest thoughts, feelings and practices of ancient civilizations, through the senses of sight, hearing, smell, and touch of muscle and nerve, to contact and utilize all the great sources of energy, harmony and knowledge of eight thousand years of wisdom and tradition. We can recharge our batteries in many different ways in order to counteract the multiple stress of twentieth century living. Through the eyes, we can view the wonderful panorama of nature, with its primeval feelings of peace and beauty. Through our ears we can contact the sounds of the wonderful symphony of nature. We can smell the fragrance of flowers and grass, and breathe deeply of the pure air in the mountains or at the ocean. Through our skin we can touch the wind, the sun, the rain, and the grass. We can still go to Nature—she is still not too far away, and we can never destroy her completely. She will always be there waiting for us when we, her prodigal children, at last return to her.

Let us, along with Shakespeare, find "tongues in trees, books in the running brooks, sermons in stones, and good in everything."

The Falls of Giessbach, Switzerland Photo by Centaur

THE SIXTEENFOLD COMPASSION OF BUDDHA

More than twenty-five hundred years ago, Buddha recognized the destructive effects of negative thoughts, and the fact that nothing so greatly inhibits our individual evolution as the admittance of these disharmonious, inferior thought-currents into our minds. His vivid description of man's self-induced suffering has come down to us in the form of his immortal Sixteenfold Compassion:

My children,

The Enlightened One, because he saw Mankind drowning in the Great Sea of Birth, Death and Sorrow, and longed to save them,

For this he was moved to compassion.

Because he saw the men of the world straying in false paths, and none to guide them,

For this he was moved to compassion.

Because he saw that they lay wallowing in the mire of the Five Lusts, in dissolute abandonment,

For this he was moved to compassion.

Because he saw them still fettered to their wealth and their possessions, knowing not how to cast them aside,

For this he was moved to compassion.

Because he saw them doing evil with hand, heart and tongue, and many times receiving the bitter fruits of sin, yet ever yielding to their desires,

For this he was moved to compassion.

Because he saw that they slaked the thirst of the Five Lusts as it were, with brackish water,

For this he was moved to compassion.

Because he saw that though they longed for happiness, they made for themselves no karma of happiness; and though they hated pain, yet willingly made for themselves a karma of pain: and though they coveted the joys of Heaven, would not follow his commandments on earth,

For this he was moved to compassion.

Because he saw them afraid of birth, old age and death, yet still pursuing the works that lead to birth, old age and death,

For this he was moved to compassion.

Because he saw them consumed by the fires of pain and sorrow, yet knowing not where to seek the still waters of Samadhi,

For this he was moved to compassion.

Because he saw them living in an evil time, subjected to tyrannous kings and suffering many ills, yet heedlessly following after pleasure,

For this he was moved to compassion.

Because he saw them living in a time of wars, wounding and killing one another: and knew that for the riotous hatred that had flourished in their hearts they were doomed to pay an endless retribution,

For this he was moved to compassion.

Because many born at the time of his incarnation had heard him preach the Holy Law, yet could not receive it,

For this he was moved to compassion.

Because some had great riches which they could not bear to give away,

For this he was moved to compassion.

Because he saw the men of the world plowing their fields, sowing the seed, trafficking, huckstering, buying and selling: and at the end winning nothing but bitterness,

For this he was moved to compassion.

Because he saw men cease to do good,

For this he was moved to compassion.

Because he saw men learn to do evil,

For this he was moved to compassion.

THE ULTIMATE MEDITATION FOR INFINITE LIVING

Be Still.
Know
I am God.
—*The Essene Gospel of Peace*

"A man who worries is like an animal
obsessed and chased by a ferocious beast,
and all around him are the green meadows. . ."
—*Goethe*

In addition to the three Communions with the Lifestream, the Soundstream, and the Lightstream, for twenty or thirty minutes each day, disconnect your mind of all the worries and inferior currents of thoughts and feelings. Empty your mind completely of all the incoming traffic of thoughts of wars, misery, disease, ignorance, violence, confusion, suffering, guilt, and other imaginary evils. Imagine you are lying relaxed in a green meadow, the grass sprinkled with beautiful wild flowers, looking up at the flying white clouds in the deep blue sky, your mind filled with only two words: BE STILL. . .

After you have emptied your mind of all the biological and psychological damage done by these inferior currents, your subconscious and conscious mind is free to absorb instead only sources of harmony and energy.

Now dedicate yourself to an hour or so of reading great books, classic masterpieces to fill with harmonious thoughts and feelings the vacuum in your mind, after you have eliminated its disharmonious contents by this daily meditation. By so doing, there will be no room for negative, inferior currents to re-enter.

Something great, spiritual and religious occurs when we sit down in silence, open a book and through the process of optical refraction certain little figures on its pages open our minds to a flow of knowledge which, like a broad, majestic river, comes pouring into our consciousness from far, remote ages, civilizations and cultures. Thoughts accumulated and condensed in these mysterious pages enter our minds. It is a mystic communion with the greatest power in the universe—Thought.

When we read a great masterpiece we hold communion with that very same power which has made Man the dominant race on Earth. Since his appearance upon this planet Man has become Homo Sapiens—Thinking Man—and continues the great work of the Creation. His dominance is due to the development of something which belongs to no other living being, no other race upon this planet: the mysterious function of Intelligence. This mysterious act of thinking which has made Man into Homo Sapiens and is his title to nobility is also the Key to the Universe.

Great books are batteries of thought, accumulators of specific energies, and reading is a communion with this force, this greatest power in the Universe—Thought. And everyone who possesses this great power of thinking and absorbing knowledge is a true collaborator in the great work of the Creation. Creation is not a mysterious act of the past but continues even unto today and will continue so long as the power to think will exist.

But we must be selective in choosing which books to read. Were we able to read a dozen books every second, we could not read more than an insignificant fragment of the existing storehouse of noble and valuable books. Time wasted in reading an inferior book is lost forever and can never be re-utilized to read a superior work. Therefore, whenever we read a mediocre book we sacrifice a great masterpiece which must remain unread because of the limitedness of our time. Before reading a book we should never forget that the reading of one involves the sacrifice of another which must remain forever unread; that "life is short, Art eternal." *Vita brevis, Ars longa.*

It is not enough to read a book, we must penetrate into its spirit, assimilate all its basic thoughts, coordinate its knowledge with that culture which we have already acquired and, finally, retain it so that it becomes an organic living part of our memory to be utilized at will whenever needed. One of my books, *The Art of Study: the Sorbonne Method,** can be a great help in mastering these techniques.

Zarathustra

*This book and all the others mentioned in this volume are available from the International Biogenic Society.

In my book, *Books, Our Eternal Companions,* I have listed many of the greatest masterpieces of universal literature as a guide to study. The following list is extremely condensed and by no means is meant to represent all the treasures available to us, as the twentieth century heirs of eight thousand years of culture.

The Vedas
The Upanishads
The Bible
The Zend Avesta of Zarathustra
The Teachings of Buddha
The Essene Gospel of Peace
The Tao Te Ching of Lao Tzu
The Meditations of Marcus Aurelius
Epictetus: Discourses and Encheiridion
Poems of Kabir, translated by Tagore
Writings of Plato and Aristotle
Spinoza: Tractatus, Ethics
Plutarch: Lives
Works of Homer and Horace
Plays of Sophocles, Euripides, Aristophanes
Complete Works of Shakespeare
Francis Bacon: Essays, Novum Organum
Goethe: Faust
Voltaire: Candide, Zadig, Micromegas
Dante: Divina Comedia
Tolstoy: War and Peace, Anna Karenina, On Life
Dostoyevsky: The Brothers Karamazov, The Idiot
Gorky: The Mother
Plays of Ibsen and Chekov
Poems of Shelley, Poe, Whitman, Browning
Plays of Molière
Writings of Victor Hugo and Balzac
Romain Rolland: Jean Christophe
Hermann Hesse: Magister Ludi, Siddhartha, Demian
Collected Poems and Plays of Rabindranath Tagore

Poems of Baudelaire, Rimbaud, Ady
Montaigne: Essays
Emerson: Essays
Writings of Hippocrates
Writings of Galileo, Newton, Darwin, Einstein
Works of Thomas Jefferson, Thomas Paine
Benjamin Franklin: Autobiography
Thoreau: Walden
Knut Hamsun: Pan, Hunger

The Earth Spirit Appears
(Drawing by Goethe)

Three Flashes of Lightning in the Gathering Storm

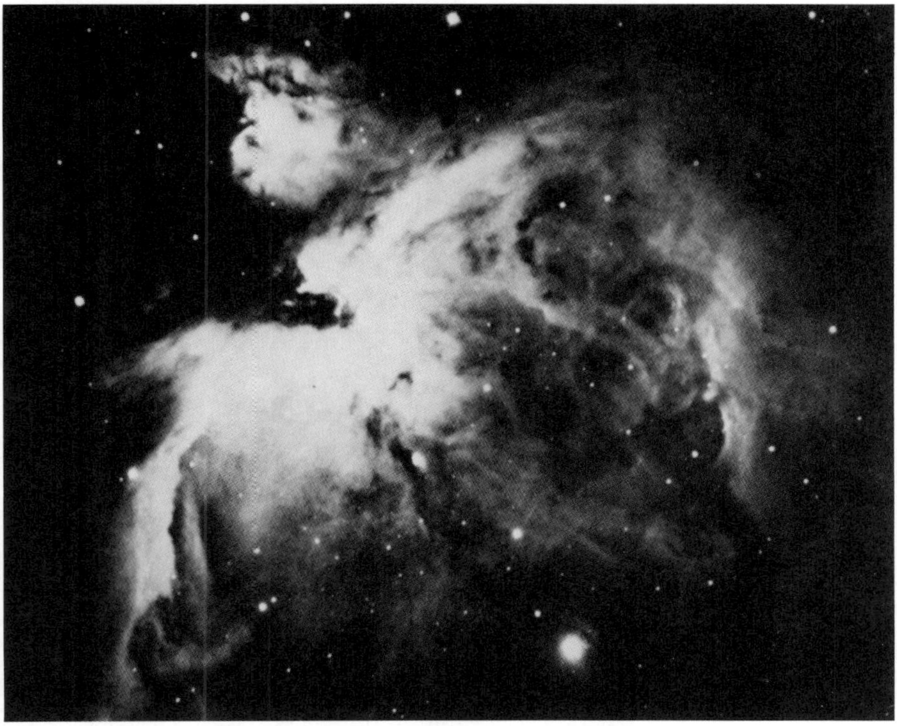

THE ESSENE COMMUNIONS WITH THE INFINITE

THE LIFESTREAM

THE SOUNDSTREAM

THE LIGHTSTREAM

I

THE MORNING COMMUNION WITH THE TREE OF LIFE

"Gray are all the theories,
But Green is the Tree of Life."
—*Goethe*

"God is the God of the living . . ."
—*The Essene Gospel of Peace*

"The Tree is the Law Itself."
—*Zarathustra*

THE COMMUNION
Brother Tree. . . Brother Life

You are floating through the green foliage of a beautiful forest, the sun filtering through the lacy web of leaves, a carpet of velvet green beneath you. You pause before a tall, majestic tree and embrace it. You feel the life-forces of the tree flowing through your whole body. As you breathe deeply, repeat to yourself the word "Life." Soon you will feel the tingling of the life-forces of the tree in your spine, and you can also direct these forces to the areas of your body which need healing. At the end of this communion, say "Thank you, Brother Tree."

THE PHILOSOPHY
The Lifestream

"In the country
It seems as if every tree
Said to me: Holy! Holy!
Who can give complete expression
To the ecstasy of the woods!
Almighty One,
In the woods I am blessed.
Happy every one in the woods.
Every tree speaks through Thee.
O God! What glory in the woodland.
On the heights is Peace—Peace to serve Him. . ."

—*Beethoven*

All the cosmic and natural laws are written in trees, in flowers, in grass, in the shining of the sun, in the flow of rivers, in rains and in wind, in the singing of the birds, in the work of the insects and in all the beauties of nature. The great book of nature, if we learn its alphabet and if we can read it, will reveal to us all the laws and secrets of nature. This superior school does not give us degrees and diplomas, but gives us knowledge, an immense source of energy and harmony. It teaches us the most fundamental laws of life. It gives us freedom of body, economic and moral liberty, and the profound values of a rich inner life.

Trees are the brothers of man. We spring from the same sources of life and are nourished by the same energies of nature. Without trees and grass, we could not exist on this planet. Trees protect the earth from excess of sun; their roots bind the humus and give it cohesion in the face of winds and hurricanes; their trunks and branches break the force of the violence of the storms, one tree protecting another and at the same time protecting the surface of the earth; they protect the earth from the dissolving forces of the rains; they take up precious elements from the earth's interior and carry them up into their leaves and fruits where they are transformed into organic form and made more perfect by the addition of various radiations and elements of the atmosphere. Then the trees give back these elements to us, their brothers, in the shape of fruits and leaves. They renew the purity of our air with oxygen, and fill our souls with the esthetic harmony of their physical beauty.

For hundreds of thousands of years, man lived according to his organovegetative system, in harmony with grass and trees. Our cerebrospinal system, the seat of our conscious activities, is only a few millennia old. Although our cerebrospinal nervous system created the great values of our cultures and civilizations, it created also an increasing complexity of life with resultant stress, conflicts, and tensions. Therefore, it is essential for us, from time to time, to refresh ourselves in the stressless, lost world of our organovegetative life, disconnecting temporarily our cerebrospinal functions, and reestablishing our unity with primeval vegetative life, especially the trees and the grass—a life without negative currents of thoughts and feelings, an existence in complete harmony with the laws of life. And, according to Zarathustra, the Tree is the Law itself.

At present the organovegetative system is in harmony with the laws of nature, while the cerebrospinal system is out of harmony with it, so the synthesis of man in the future must have both systems in harmony with the laws of nature. In order to get rid of the great difference between the two systems, the organovegetative and the cerebrospinal systems, man must reach the state of being able to awaken now latent and dormant capacities. If these latent energies are aroused then the cerebrospinal system will also come into harmony with the laws of nature, like our unconscious actions, and in this way the natural forces and laws will become conscious in man himself. Hitherto, and at the present time particularly, the natural forces and laws are unconscious in the human organism. The virtualization of these latent higher capacities will also bring the organovegetative system into approximation with the perfected cerebrospinal system. It was to this end that the Essene Communions were practiced thousands of years ago, and may be practiced today with the same sublime goal in mind.

Just as human society cannot evolve harmoniously until it can solve its inner contradictions and antagonisms, no more can the human organism. There is a state of the human organism when the latter is in complete harmony with the laws of nature. This occurs when the cerebrospinal system is at rest and is not working, when the organovegetative system governs all the activities of the body. This is the case when the human organism is unconscious during sleep. Some hours of sleep in harmony with the natural laws and forces are capable of entirely refreshing and revitalizing the human organism. Time spent in harmony with the natural laws and forces represent a very great source of harmony and energy for the human organism. Our conscious efforts to merge ourselves with the Cosmic Ocean of Life, Sound, and Light, through the morning, noon, and evening communions, will gradually bring us closer and closer to this perfect harmony of the millions-of-years-old organovegetative system and the younger cerebrospinal system, finally achieving Oneness with the Infinite.

We need harmony in our thoughts no less than in our diet or in physical activity. If someone has an inharmonious, ill-intentioned thought against others, he is punished for it in the moment of its creation, for the inharmonious thoughts immediately have a destructive influence on the human glands. So in the form of glandular activity the poisonous secretion always represents an immediate physiological sanction and punishment of the sin. There exists a superior tribunal of the cosmic and natural forces which punishes offenses at the moment when they are committed. We can deceive the laws of human society; we can commit crimes and escape punishment; but we can never commit a single sin against the laws of nature for which we are not punished immediately in the moment of its commission.

In the human organism there is a miraculous collaboration of every cell and organ; they display a great collective solidarity. The heart labors not only for itself, but for the whole organism, and gives blood to all the organs. Likewise the lungs introduce oxygen not only for themselves, but for other organs, and the digestive system sends chemical energies to every part of the body. So the human organism represents the natural laws through the organovegetative system; it represents altruism, collective solidarity. This law is written large in the human organism, if we are able to read in the great book of nature. I do not consider altruism and solidarity as just moral, theoretical considerations, but as an iron law of the natural forces, as a dynamic reality in the human organism and in nature. It is a law and a force, and when we act against that law and force, then immediately and inevitably comes the punishment. If we do not follow the example of solidarity and altruism represented by the human organism (and also by the trees and the grass and by everything which is in harmony with the natural laws and forces), then we are punished. There is, so to speak, a great higher tribunal of the natural and cosmic forces which formulates and issues higher laws which are valid for everything which exists in nature or in man himself; this power punishes and rewards immediately those who are in disharmony or harmony with the biological laws and forces of life.

Therefore, we shall attribute a great importance to psychic harmony; to the elimination of the internal contradictions of man, as an inescapable precondition of health, longevity, and harmony. In addition to the material forces of nature, we also use all the forces of nature which are not material, but form the superstructure of matter, for they react upon the material foundation and are capable of improving or impairing it. So not only do we take our therapeutic forces from the sphere of nature, but also from the sphere of human consciousness. For it is not possible to solve the internal contradictions of the human organism simply on a material basis. As man's name of Homo Sapiens indicates, the essence of man lies in his psychic forces, which give him his superiority over all the other creatures of the earth, provided that his forces and thoughts are used in a superior way, for if not, they can lead to destruction instead of progress. It is also a natural law that the higher and finer forms of energy are always stronger than the inferior and grosser forms. Finer radiations and forces always traverse the coarser radiations, having a greater force and penetration. For instance, there is a force of rigid matter, a force of cohesion, a physical, material force; there are electrical and magnetic forces which are finer forces, having a great power of penetration, and in spite of the fact that they are not so observable as the forces of cohesion of a solid body, nevertheless they penetrate and traverse the solid body. Similarly currents which are even higher and finer penetrate the electrical currents, and so on. There is an infinite hierarchy of the forces of nature which succeed one another in force, intensity and superiority. And the finest and strongest energy of the human organism is the force of thought.

THE LIFESTREAM IN OUR DAILY LIVES

Not only our way of thinking, but also our way of living and eating have a profound influence on our spiritual evolution. We should eat only foods which are whole, pure and natural, as close to their original state as possible, and not contaminated with chemicals and artificial processing. In *The Essene Way–Biogenic Living* and *The Chemistry of Youth*, all the details of the ideal way of eating are enumerated, but it is also possible to describe this ideal in a few words from the immortal *Essene Gospel of Peace:*

> *Prepare and eat all fruits of trees, and all grasses of the fields, and all milk of beasts good for eating. For all these are fed and ripened by the fire of life; all are the gift of the angels of our Earthly Mother.*

The living angels of the living God serve only living men. For God is the God of the living, and not the God of the dead.

So eat always from the table of God: the fruits of the trees, the grain and grasses of the field, the milk of beasts, and the honey of bees.

For I tell you truly, the God of the living is richer than all the rich of the earth, and his abundant table is richer than the richest table of feasting of all the rich upon the earth. Eat, therefore, all your life at the table of our Earthly Mother, and you will never see want. And when you eat at her table, eat all things even as they are found on the table of the Earthly Mother. Cook not, neither mix all things one with another. And you will never see disease, for such finds favor in the eyes of the Lord.

Then your days will be long upon the earth, for this is pleasing in the eyes of the Lord. Eat always when the table of God is served before you, and eat always of that which you find upon the table of God. For I tell you truly, God knows well what your body needs, and when it needs. For the power of God's angels enters into you with the living food which the Lord gives you from his royal table.

For I tell you truly, the body of the Sons of Man is turned into a temple, and their inwards into an altar, if they do the commandments of God. Wherefore, put naught upon the altar of the Lord when your spirit is vexed, neither think upon any one with anger in the temple of God. And enter only into the Lord's sanctuary when you feel in yourselves the call of his angels, for all that you eat in sorrow, or in anger, or without desire, becomes a poison in your body.

Rejoice, therefore, always with God's angels at their royal table, for this is pleasing to the heart of the Lord. And your life will be long upon the earth, for the most precious of God's servants will serve you all your days: the angel of Joy.

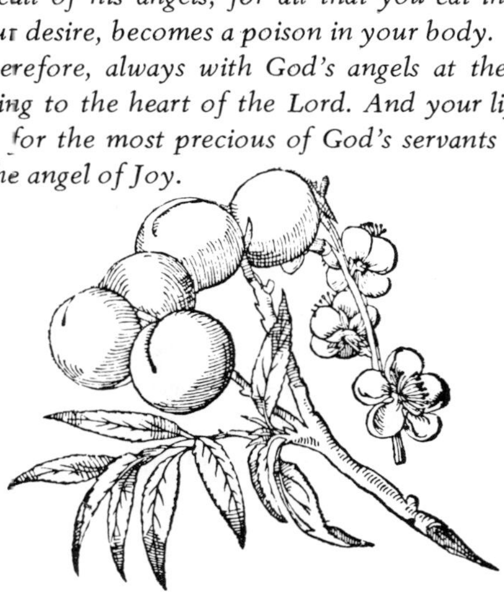

II

THE NOON COMMUNION WITH THE COSMIC OCEAN OF SOUND

"In the beginning was the Sound,
and the Sound was with God,
and the Sound was God. . ."
–*The Essene Gospel of Peace*

"Is there any wise man who will listen
To that solemn music which arises in the sky?
He, the Source of all Music,
He is the Primal Word. . ."
–*Kabir*

"I hear the melody of His flute,
And I cannot contain myself. . .
The sky roars and the lightning flashes,
The waves arise in my heart. . .
Where the rhythm of the world rises and falls,
Thither my heart has reached:
His work and His rest are filled with music. . ."
–*Kabir*

THE COMMUNION
Mother Sound

You are lying relaxed in a mountain stream, the water flowing cool around you in a direction from head to foot. On both sides of the stream are banks of flowers and grass, and in the distance shrubs and trees, warm sunlight filtering through the green leaves. Breathe deeply, repeat to yourself the words "Sound, Stream," and concentrate only on the rushing, soothing sound of the flowing waters of the swift stream. Let the sound penetrate your mind and body until you hear in your inner ear the "Music of the Spheres" of Pythagoras. At the end of this communion, say "Thank you, Mother Earth, for your sounds, healing my body and mind, and lifting me into the presence of the Eternal and Infinite Soundstream."

THE PHILOSOPHY
The Soundstream

> "Now, what is music? Music is a strange thing. I would almost say it is miracle. For it stands halfway between thought and phenomenon, between spirit and matter, a sort of nebulous mediator, like and unlike each of the things it mediates—spirit that requires manifestation in time and matter that can do without space.
>
> "We do not know what music is."
>
> —*Heine*

According to Pythagoras, we are bathing in the Music of the Spheres from the moment of our birth, but because of that very fact, we are not aware of this holy Sound which fills our inner ears. Only if we were to suddenly stop hearing it, would we be aware of it, through its loss. When we practice this Ultimate Communion with Mother Sound, we gradually come to an awareness of this Eternal Music which is one of the paths of the Infinite. *"In the beginning was the Sound, and the Sound was God. . ."*

THE SOUNDSTREAM IN OUR DAILY LIVES

You should listen each day to the sublime music of the Great Masters, or beautiful peasant songs from the purity and simplicity of authentic folk music. You must from time to time disconnect yourself from industrial noise, traffic noise, radio noise, television noise, human noise, etc., and expose your body to the healing and your mind to the uplifting sound streams of the immortal music of the Great Masters. Beethoven, Bach, Mozart—these were angels sent from the Cosmic Ocean of Life to build for suffering humanity a bridge to the Heavenly Father. For the musical works of these Great Masters are also Holy Scripture; and the *Missa Solemnis* of Beethoven is as profound a spiritual revelation as the Sermon on the Mount, of Jesus; or the

Sermon at Deer Park in Benares, of Buddha. One does not need musical training or education in order to appreciate these masterpieces. They speak directly to the soul, they enter our spiritual soundstream, as pure honey enters into the bloodstream—nourishing, uplifting, bringing divine energy to the soul, mind, and body.

Plato said, "Music is a moral law. It gives a soul to the universe, wings to the mind, flights to the imagination, a charm to sadness, gaiety and life to everything. It is the essence of order, and leads to all that is good, just, and beautiful, of which it is the invisible, but nevertheless dazzling, passionate, and eternal form."

Here is a very condensed list of some of the greatest musical masterpieces of all time. Even mentioning only three composers, it is by no means a complete list, but it may serve as a first step in the direction of acquiring a library of immortal music—an earthly reflection of the Great Cosmic Soundstream which we will hear, little by little, as we practice the Noon Communion with the Music of the Spheres.

Ludwig van Beethoven

- Missa Solemnis
- The Nine Symphonies
- The Thirty-Two Piano Sonatas
- The Five Piano Concerti
- The String Quartets (especially the Late Ones)

Johann Sebastian Bach

- The Mass in B Minor
- The St. Matthew Passion
- All Organ and Harpsichord Music*
- The Musical Offering
- The Art of the Fugue

*Glenn Gould, one of the most gifted pianists of this century, has recorded all of Bach's harpsichord music on the piano, in exquisitely sensitive interpretation.

Wolfgang Amadeus Mozart
The Quartets
The Piano Concerti
All Other Piano Music
All Sacred Music
The Great C Minor Mass
Requiem

BACH AND MOZART

Bach, the Heavenly Mathematician—and Mozart, the Divine Orpheus, were in their personalities and ways of living, very different. And at first hearing, their music will seem also dissimilar to the listener. But the quality of genius is identical, and to listen to their sacred music is a spiritual banquet of sound. In the magnificent *Mass in B Minor* of Bach, and in Mozart's *Requiem* and the *Great Mass in C Minor,* we ascend through the heights in marvelously ordered stages, as divine musical equations reveal the Cosmic Order. For one who hears Bach's *Mass in B Minor* for the first time, these words may provide a kind of introduction:

". . . he reinforces and lifts and rounds out his edifice of notes far above the church into a starry space full of nobly perfect systems, as though God had gone to sleep and handed over to him His staff and mantle. He makes the lightning play from towering clouds and throws open serene sunny spaces, he triumphantly guides forth planets and suns, he rests relaxed at high noon and at the proper hour elicits the cool showers of evening. And he ends in splendor and power like the setting sun and, as he falls silent, leaves the world full of glory and of soul. . ."

from the Credo of the Missa Solemnis.

LUDWIG VAN BEETHOVEN
and the Missa Solemnis

> ". . . the work of Beethoven parallels the pillars of smoke and fire which led the Israelites through the desert, a pillar of smoke to lead us by day, and a pillar of fire to light the night, so that we may march ahead both day and night. His darkness and his light equally trace for us the road we must follow; both the one and the other are a perpetual commandment, an infallible revelation."
>
> —*Franz Liszt*

Of all the great works of the great Masters, there is none which ascends higher to the infinite realms than the *Missa Solemnis* of Beethoven, that awesome work which towers over all other music as the ceiling of the Sistine Chapel dwarfs all the symbols of pomp and power passing beneath it. In the *Credo,* after the words "et sepultus est"—he was buried—have been sung in whispered tones of unbearable sorrow and longing, there is a pause in the music—a pause of agonizing suspense, not a soul breathing in heaven or earth. And then—the miracle happens—the tenor voices exultantly proclaim "Et resurrexit! Et ascendit in coelum!" Surely there has never been enacted a greater moment of glory, of blinding ecstasy, of absolute faith in the ultimate triumph of the spirit of man, than in that clarion call of joyful reunion with God: "And he arose! And he ascended to heaven!"

The clenched fist shaking in anger as Beethoven died was not directed to God. That gesture spoke of the frustration of a towering genius who was also a man of compassion, who all his life wanted desperately to bring the sacred fire of his music to his fellow man, but knew at the end that most had neither the maturity nor the strength to enter into his world. Beethoven once said that anyone who truly understood his music would never know sadness. But the day is still far in the future when we will truly understand those mysterious combinations of notes and sounds which were somehow transmitted to our world, through the genius of the modern Prometheus, Beethoven, from vast unknown regions of unutterable splendor.

Here are the words of Beethoven:

"When I open my eyes I must sigh, for what I look upon is contrary to my religion, and I must despise the world which never divines that music is a greater revelation than the whole of wisdom and philosophy; music is the wine that incites us to new creation, and I am the Bacchus who presses this glorious wine for mankind and grants them drunkenness of spirit; when they are sober again they will have fished up much which they may take with them to dry land. . . nor am I at all anxious about the fate of my music; its fate cannot be other than happy; whoever succeeds in grasping it shall be absolved from all the misery that bows down other men."

And so Beethoven broke his chains, with a wrathful fist, and with the shattering splendor of the *Credo* of the *Missa Solemnis.* But he left with us the fire. At what terrible cost did he leave us that fire! And through the searing majesty of those flames of sound, we may one day come to understand who he really was.

These words were kept always on his desk, framed under glass:

I am that which is.
I am all what is, what was, what will be;
No mortal man has lifted my veil.

Beethoven died about ten days after beginning this sketch for a Tenth Symphony.

Day and night, the chorus of music fills the heavens,
And a mighty sound breaks forth in song.
There the Unstruck Music sounds of itself,
And the streams of Light flow in all directions.
What a wonderful lotus it is,
That blooms at the heart of
The spinning wheel of the universe!
Music is all around it,
And all space is radiant with Light.
There the whole sky is filled with sound,
And there that music is made without fingers
And without strings.
The middle region of the sky,
Wherein the spirit dwelleth,
Is radiant with the music of Light. . .
Dance, my heart!
Dance today with joy—
Life and Death dance to the rhythm of this music.
The Hills and the Sea and the Earth dance.
Whether I rise or sit down, I can never forget Him,
For the rhythm of His music beats in my ears.

—Kabir

III

THE EVENING COMMUNION WITH THE COSMIC OCEAN OF LIGHT

"And God said, Let there be Light. . .
And there was Light."

—*Genesis*

Et Lux in tenebris lucet. . .
And the Light shineth in the darkness. . .

". . . the Eternal Light to shine
before the Children of Israel. . ."

—*Old Testament*

THE COMMUNION
Father Light

You are floating, floating on the still, untroubled surface of a lake, surrounded by mountains and forest. Above you is the sky filled with stars. It is night, but darkness is dispelled by the thousands and thousands of brilliant diamond-flashes of stars, shining, streaming with Cosmic Light. Breathe deeply, repeat to yourself the words "Light, Light," and concentrate only on the star-filled sky until you will see all kinds of light formations behind your closed eyes. Visualize yourself floating in the Light of the Stars, floating upward, ascending to become one of the Angels of the Heavenly Father.

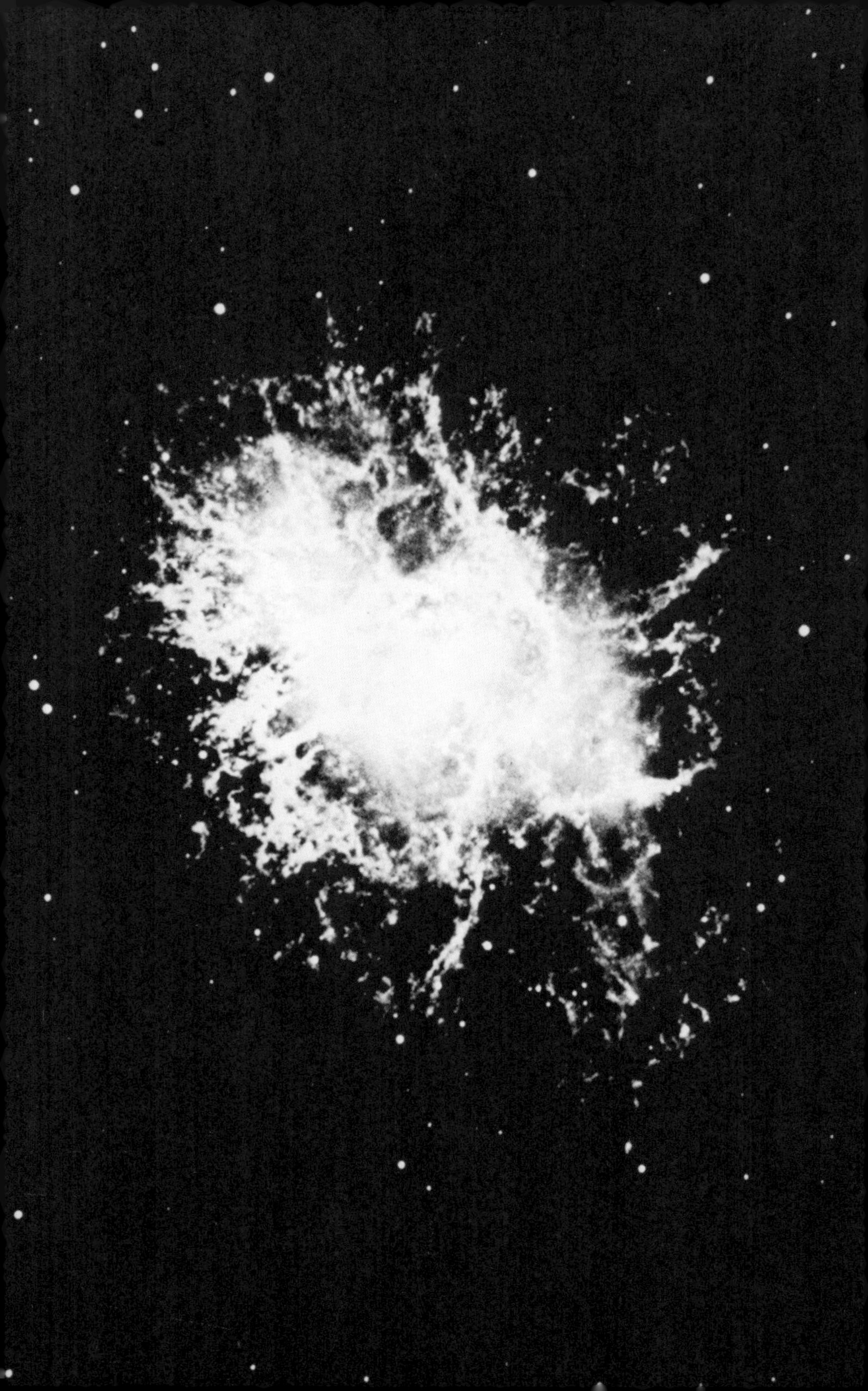

THE PHILOSOPHY
The Lightstream

"Two things fill my soul
with ever new and increasing
wonder and reverence:
the starry sky above me,
and the moral law within me."
—*Kant*

In concentrating on the Lightstream, the eternal brilliance of the Cosmic Ocean of Light, we gradually set the stage for the final step that will unite us with the Heavenly Father, reaching complete union with the totality of all laws, the One Law. By submerging ourselves in an Ocean of Light, and concentrating on the star-filled sky before going to sleep, we ensure that through the night our subconscious mind will be in contact with the vast storehouse of superior cosmic forces, making it possible for sleep to become a source of deepest knowledge.

In this communion, images are much better than words. We learn from the study of Archeosophy* that the way of the ancients in expressing ideas, in using images, pictographs and symbols, is nearer to reality, more vital and living, than are abstract letters and words and sentences. Pictures, no matter how primitive, awaken responses. Even a horizontal line produces a feeling, suggesting that which is static and at rest. A vertical line appears dynamic, in movement. The color red gives the impression of ebullition, excitement, whereas blue produces a feeling of coolness, of that which is soothing and lofty. These reactions are universal. From time imme-

*See *Archeosophy, a New Science,* by Edmond Bordeaux Szekely.

morial geometric figures and colors have awakened responses, stirred emotions, stimulated ideas. However high a degree of exactitude our contemporary civilization may have reached in the imparting of analytical knowledge, it is far more inadequate than the ancient symbols in the conveyance of world pictures and wisdom. Ancient symbolism shortens the path to the communication of intuitions. And Archeosophy crosses a borderline, abandoning the apparently solid ground of alphabets and words and alphabetized knowledge, entering a completely different world where there are none of the familiar landmarks of letters, only symbols. Symbols do more than express inner intuitions of the creator of the symbol. A dynamic emotional and intellectual unity is established between the author or creator, and the individual who contemplates the symbol. There is a direct current of emotions and thoughts between them.

This is why when we contemplate the star-filled sky and submerge ourselves in the Infinite and Eternal Lightstream before going to sleep, we make direct contact with the Cosmic Ocean of Eternal Life.

THE LIGHTSTREAM IN OUR DAILY LIVES

During the day, we can also enhance our communion with the Visible Eternal Kaleidoscope by letting our eyes feast on the beauties of nature and the great works of art which celebrate nature. The paintings and sculptures of Leonardo da Vinci and Michelangelo, all the great classic masterpieces of art, should occupy our visual attention as much as possible, especially when we are not physically able to enjoy the sights of nature. We should make a collection of the most beautiful landscape paintings (Dürer, El Greco, Brueghel, Rembrandt, Turner, etc.) which we can enjoy at any time, in order to penetrate sublime nature through the eyes. And of course, when we have the good fortune to be in the country, in the mountains, in the forest, by the ocean, we should enjoy to the fullest the spiritual banquet given us by Nature, and not allow any inferior currents of thought or feeling intrude.

Crossing the Brook, by J.M.W. Turner. National Gallery, London.

WE ARE ONE WITH THE UNIVERSE

We must always be aware of our essential unity with not only the stars, but also the music of the spheres, the life force in the trees and grass, in fact with all of Creation. We are one atom in this infinite cosmic life, and as part of the infinite, we too possess life eternal. We are eternal by our ability to absorb all sources of energy, harmony, and knowledge in our earthly, cosmic, and spiritual environment. We are the product of an endless evolution bearing us towards ever-higher forms of existence. We enjoy perfect unity with the universe, not duality. Our bodies are made of the same electrons, atoms, and molecules as constitute the celestial bodies of the galactic and ultra-galactic systems. Our skeleton is composed of the same minerals as constitute the different geological strata of the earth. The circulation of our blood reproduces in miniature the circulation of water on the earth's surface. This unity with nature and the universe is not a mere philosophical theory, but a living reality in which we live permanently through intense experience of body and soul. *And the Light shineth in the darkness. . .*

THE LIGHTSTREAM OF THE HEAVENLY FATHER AS DESCRIBED BY ZARATHUSTRA IN THE ZEND AVESTA

(more than six thousand years ago)

We invoke
The bright and glorious Stars,
That from the shining East
Move along their winding course,
Along the path made by the gods,
Along the way appointed for them
At the Will of the Ahuras.
We sacrifice unto
The bright and glorious Stars,
Whose rising is watched by men
Who live on the fruits of the Year,
By men of deep understanding.
We invoke
The bright and glorious Stars,
That wash away
All things of fear
And bring health
Unto all Creations.
The bright and glorious Stars
Which Ahura Mazda hath established
In Infinite Space:
We sacrifice unto
The bright and glorious Stars
To which Ahura Mazda
Hath given a thousand senses,
That have within themselves
The seed of Life
And of Water,
That have in them
The seeds of Plants.
The white, shining,

Far seen Stars!
The piercing, Health-bringing,
High, far-piercing–
With their shining rays;
For their Brightness and Glory
I shall offer unto them
A sacrifice.
The bright and glorious Stars
That fly towards the Sea
As swiftly as the arrow darteth
Through Heavenly Space.
We invoke
The bright and glorious Stars,
That stand out beautiful,
Spreading Comfort and Joy
On the fertile Countries,
Communing within Themselves. . .

The moon shines in my body,
But my blind eyes cannot see it.
The moon is within me,
And so is the sun.
He Himself is the sun,
The light, and the lighted.
The light of the sun, the moon,
And the stars, shines bright:
The melody of love swells forth,
And the chorus of music fills the heavens.
My Beloved One gleams
Like the lightning flash in the sky.
They have sung of Him
As infinite and unattainable;
But I in my meditations
Have seen Him without sight.
In the temple of the sky
The light of the sun and moon are shining;
Still your mind to silence before that splendor.
There it is ever moonlight and never dark;
And who speaks of one sun only?
That land is illuminate with the rays
Of a million suns ablaze with light.
Hark to the unstruck bells and drums!
Rains pour down without water,
And the rivers are streams of Light . . .

—Kabir

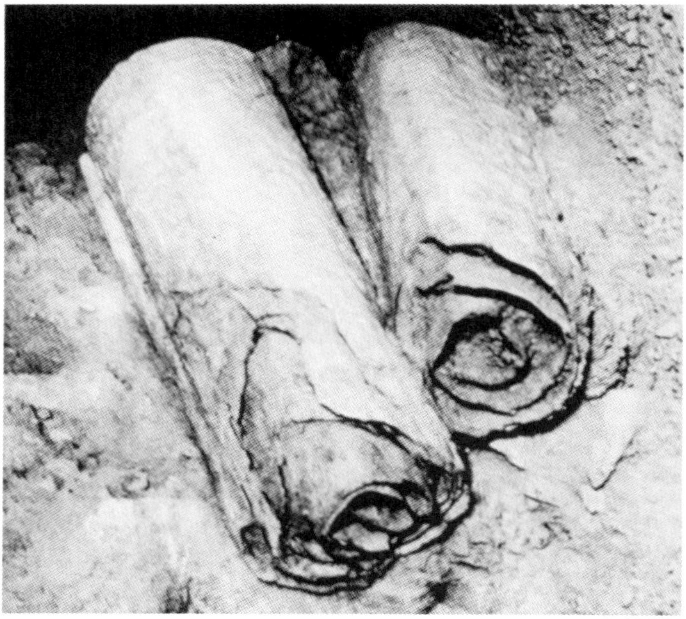

THE LIFESTREAM, THE SOUNDSTREAM, THE LIGHTSTREAM AS REFLECTED IN THE GREAT TEACHINGS OF 8000 YEARS

We should keep before us constantly the greatest wisdom of the last 8000 years, those spiritual teachings in their original purity and simplicity which will instill in our lives only the highest standards of ethical conduct. We must never make the fatal error of becoming one-sided, for those who believe they are in possession of the only and absolute truth have committed a kind of intellectual suicide. By studying the teachings of all the great masters of the past (but making sure they are the original teachings, not the "commentaries on the commentaries on the commentaries"), we ensure that our individual evolution will progress in an optimal, omnilateral way.

Here are a few gleanings from the greatest wisdom of the last 8000 years, each one representing a link in that infinite chain of spiritual revelations which have been given to man, usually at a time of darkness and need, all through our history of existence on our planet.

FROM THE UPANISHADS

"To love is to go on teaching forever
without becoming weary."

May quietness descend upon my limbs,
My speech, my breath, my eyes, my ears:
May all my senses grow clear and strong.
May Brahman show himself to me.
Never may I deny Brahman, nor Brahman me.
I with him and he with me—may we walk always together.
May there be revealed to me, who am devoted to Brahman,
The whole truth of the Upanishads.
AUM. . . Peace, Peace, Peace. . . be always everywhere.
O thou self-luminous Brahman, remove the veil of ignorance
From before me, that I may behold thy light.
Do thou reveal to me the spirit of the scriptures.
May the truth of the scriptures be ever present to me.
May I seek day and night to realize what I learn from the sages.
May I speak the truth of Brahman.
May I speak the truth.
May it protect me.
AUM. . . Peace, Peace, Peace. . . be always everywhere.
May Brahman protect us,
May he guide us,
May he give us power and right understanding.
May love and peace be with us all.
AUM. . . Peace, Peace, Peace. . . be always everywhere.
Lead me from the unreal to the real.
Lead me from darkness to light.
Lead me from death to immortality.
AUM. . . Peace, Peace, Peace. . . be always everywhere.
With our ears may we hear what is good.
With our eyes may we behold thy truth.
Restful in body, may we who worship thee find peace.
Filled with Brahman are the things we see,
Filled with Brahman are the things we see not,
From out of Brahman flows all that is:
From Brahman all. . . yet is he still the same.
AUM. . . Peace, Peace, Peace. . . be always everywhere.

THE HYMN OF CREATION

from the Rig-Veda

Then was not non-existent nor existent: there was no realm of air, no sky beyond it.

What covered in, and where? And what gave shelter? Was water there, unfathomed depth of water?

Death was not then nor was there aught immortal: no sign was there, the days and nights divided.

That one thing, breathless, breathed by its own nature: apart from it was nothing whatsoever.

Darkness there was: at first concealed in darkness, this All was indiscriminated chaos.

All that existed then was void and formless: by the great power of warmth was born that unit.

Thereafter rose desire in the beginning. . . Desire: the primal seed and germ of spirit.

Sages who searched with their hearts' thought discovered the existent's kinship with the non-existent.

Transversely was their severing line extended: what was above it then, and what below it?

There were begetters, there were mighty forces, free action here and energy up yonder.

Who verily knows and who can here declare it, whence it was born and whence comes this creation?

The gods are later than this world's production. Who knows, then, when it first came into being?

He, the first origin of this creation, whether he formed it all or did not form it,

Whose eye controls this world in highest heaven, he verily knows it. . . or perhaps he knows not.

FROM THE TEACHINGS OF BUDDHA

"Strive Incessantly."

—the last words of Buddha

We are surrounded by an Ocean of Suffering, the Samsara. There is nothing more important and urgent than to remedy suffering in and around us through the understanding of the Four Noble Truths and the following of the Noble Eightfold Path. The two causes of suffering are Ignorance *(Avidja)* and Thirst *(Tanha)* for the wrong pleasures, for which we have to pay two very high prices: our health and our peace of mind.

Health is the greatest blessing, contentment the best possession, a true friend the nearest of kin.

Attach yourself to the wise, to the intelligent, to the learned. Sweet is the solitude and peace of mind. Sweet is to be free from fear and desire. Hold fast to the truth as your refuge.*

THE HYMN OF AKHENATEN

from Ancient Egypt

O Living Aten, beginning of life! Though Thou art afar, Thy rays are on earth: though Thou art on high, Thy footprints are the day.

How manifold are all Thy works! They are hidden from before us, O Thou sole god whose powers no other possesseth.

Thou makest the beauty of form through Thyself alone.

Thou art in my heart; there is none other that knoweth Thee save Thy son Akhenaten.

For Thou art duration, beyond Thy mere limbs; by Thee man liveth.

Thy world is in Thy hand, even as Thou hast made them.

*From *The Living Buddha,* by Edmond Bordeaux Szekely.

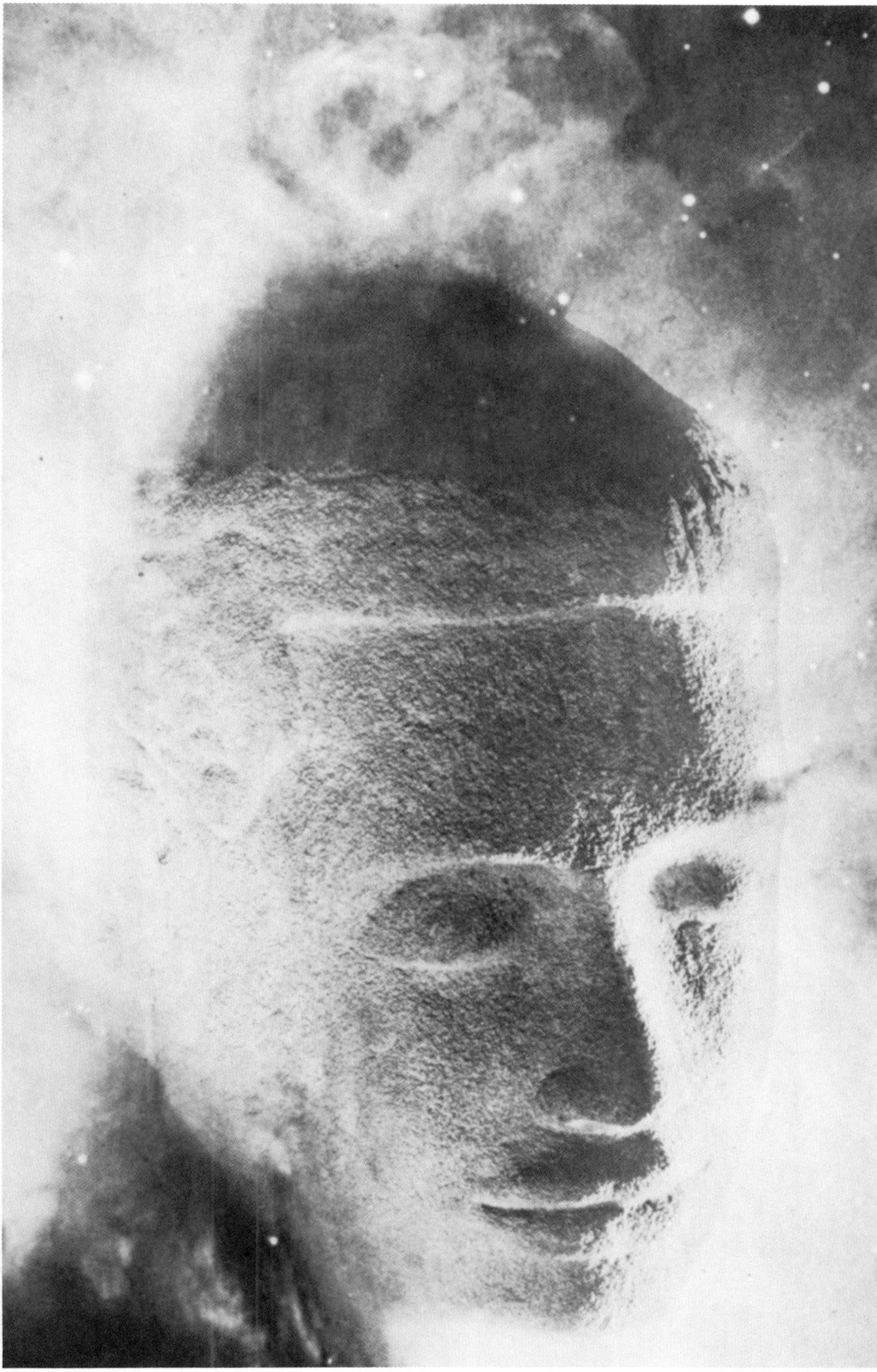

FROM THE TAO TE CHING OF LAO TZU

There was something mysteriously formed, born before heaven and earth. . . in the silence and void, standing unchanging and alone.

I do not know its name. . . call it Tao.

Man follows the ways of earth, earth follows the ways of heaven, heaven follows the ways of Tao, Tao follows its own way.

The Tao that can be explained is not the eternal Tao. The name is not the eternal name. The nameless is the beginning of heaven and earth. The named is the mother of ten thousand things.

Desireless, one can see the Eternal. Desiring, one can see the temporal. These two spring from the same source but differ in name—this appears as darkness. Darkness within darkness—the key to all mystery.

Empty yourself of everything, and let the mind rest in peace. The ten thousand things rise and fall; they grow and bloom and then return to the source. Returning to the source is stillness, which is the way of nature.

Being at one with the Tao is eternal. And though the body disappears, the Tao will never pass away.

Man, born tender and yielding, stiffens and hardens in death. Green plants are tender and filled with sap; at their death they are withered and dry.

The stiff and unbending is the disciple of death. The gentle and yielding is the disciple of life.

The hard and strong will fall; the gentle and weak will overcome.

To know others is wisdom; to know one's self is enlightenment. To master others requires force; mastering the self needs strength.

Stay with the primeval Tao, and move with the present. To know the ancient beginning is the essence of Tao.

FROM THE ESSENE GOSPEL OF PEACE

The laws are written in everything that has life; in the grass, the trees, the rivers, mountains, birds of heaven and fish of the sea; but chiefly in man himself.

Life is the ever-living word, teaching the true laws to man.

The laws of life are in your heart and spirit, your breath and bone, your flesh and every least part of your body.

Seek the fresh air of the forests, of the fields; put off your shoes and your clothing and suffer the air to embrace all your body. Then breathe long and deeply that the angel of air may be brought within you.

Seek the angel of water; put off your shoes and your clothing and suffer the angel of water to embrace all your body.

Seek the angel of sunlight; put off your shoes and your clothing and suffer the sunlight to embrace all your body. Then breathe long and deeply that the angel of sunlight may be brought within you.

The angels of air and water and sunlight are brethren given to you that you may go always from one to the other.

In the air, the water, the earth, plants, sunbeams, in the depths and in the heights the laws of life speak to you, that you may understand the words and the will of the living God.

The foods you eat from the table of God—the fruits of trees, the grain and grasses of the field, the milk of beasts and the honey of bees—give strength and youth to the body and you will never see disease.

Your body is that which you eat and your spirit is that which you think.

Your Heavenly Father is Love.
Your Earthly Mother is Love.
The Son of Man is Love.
Love is Eternal.
Love is Stronger than Death.

THE CREATION

from the Essene Book of Creation

Without beginning, the Law creates Thought and Life.

Everything that exists, materially and immaterially, follows the Law, the One Law which is everywhere ever present, behind all, governing all.

Man punishes himself when he deviates from the Law but through his deviations he acquires experience and knowledge.

When we become conscious of the Law, work with it, understand it, it is a source of energy, knowledge, and harmony. The Law is termed God by some people.

Creation is continuous and eternal, endless in time and space, an ever present universal process, going on everywhere in the infinite universe.

Creation is continually going on in the human consciousness, creating thoughts.

All inorganic matter contains in itself the potentiality of life. The earth is not the only planet in the universe and the continuous creation of the Law goes on throughout the Cosmic Ocean of Life.

THE LAW

from the Essene Book of Creation

The Law does not exist: it is valid. The Law is formless, contains all wisdom, all power, all love, all reality. It never became; it never disappears.

The Law is everywhere present in the universe, for it is the cohesive force which binds the universe together in all its parts.

Man can no more be apart from the unity of the Cosmic Ocean than a cell in his body can be apart from his body.

Man's illusion of separateness is caused by his deviations from the Law, which impose self-created limitations in his life.

ENOCH

The esoteric goal which Moses projected for humanity is express in his words:

> *"And Enoch walked with God:*
> *and he was not; God took him."*

Few have ever comprehended the incomparable grandeur of this sentence. Moses was stating that the consciousness of Enoch was united with the Cosmic Consciousness and dissolved into it.

The realization of this state which he himself reached on Mount Nebo, has been the supreme goal Moses projected for humanity.

FROM ANCIENT ROME

EPICTETUS

The philosopher does not rely on reputation nor on wealth nor on the power of a magistrate, but on his own strength, that is, on his opinions about the things which are in our power and those which are not. For these are the only things which make men free, which make them escape from hindrance, which raise the head of those who are depressed, which make them look with steady eyes on the rich and on tyrants. And this was the gift given to the philosopher.

MARCUS AURELIUS

Men seek retreats for themselves, houses in the country, seashores, and mountains. But this is altogether a mark of the most common sort of men, for it is in thy power whenever thou shalt choose to retire into thyself. For nowhere either with more quiet or more freedom from trouble does a man retire into his own soul, particularly when he has within him such thoughts that by looking into them he is immediately in perfect tranquility; and I affirm that tranquility is nothing else than the good ordering of the mind. Constantly then give to thyself this retreat, and renew thyself.

HYMN TO THE SUN OF ST. FRANCIS

Praise be to Thee, O Lord,
For all Thy creatures,
And especially for our Brother, the Sun,
Who gives us the day,
And who shows forth Thy light.
Fair is he and radiant with great splendor:
To us he is the symbol of Thee, O Lord.

Praise be to Thee, O Lord,
For our sister, the moon,
And for the stars.
Thou hast set them clear,
Beautiful and precious in the heaven above.

Praise be to Thee, O Lord,
For our brother, the wind,
For the air and the clouds,
For the clear sky and for all weathers,
By which Thou givest life
And the means of life
To all Thy creatures.

Praise be to Thee, O Lord,
For our brother, fire,
By whom Thou givest us
Light in the darkness.
He is beautiful and bright,
Courageous and strong.

Praise be to Thee, O Lord,
For our sister, water,
Who is so useful to us,
Humble, precious, and chaste.

Praise be to Thee, O Lord,
For our mother, the earth,
Who sustains and nourishes us,
Bringing forth diverse fruits,
Flowers of many colors,
And the grass.

FROM THE DEAD SEA SCROLLS

I thank Thee, Heavenly Father,
because Thou hast put me
at a source of running streams,
at a living spring in a land of drought,
watering an eternal garden of wonders,
the Tree of Life, mystery of mysteries,
growing everlasting branches for eternal planting
to sink their roots into the stream of life
from an eternal source.
And Thou, Heavenly Father,
protect their fruits
with the angels of the day
and of the night
and with flames of eternal Light burning every way.

—from the Thanksgiving Psalms
of the Dead Sea Scrolls

I will praise Thy works
with songs of Thanksgiving
continually, from period to period,
in the circuits of the day, and in its fixed order;
with the coming of light from its source
and at the turn of evening and the outgoing of light,
at the outgoing of darkness and the coming in of day,
continually,
in all the generations of time.

—from the Thanksgiving Psalms
of the Dead Sea Scrolls

I am grateful, Heavenly Father,
for Thou hast raised me to an eternal height
and I walk in the wonders of the plain.
Thou gavest me guidance
to reach Thine eternal company
from the depths of the earth.
Thou hast purified my body
to join the army of the angels of the earth
and my spirit to reach
the congregation of the heavenly angels.
Thou gavest man eternity
to praise at dawn and dusk
Thy works and wonders
in joyful song.

—from the Thanksgiving Psalms
of the Dead Sea Scrolls

I have reached the inner vision
and through Thy spirit in me
I have heard Thy wondrous secret.
Through Thy mystic insight
Thou hast caused a spring of knowledge
to well up within me,
a fountain of power,
pouring forth living waters,
a flood of love
and of all-embracing wisdom
like the splendor of eternal Light.

—from the Book of Hymns of the Dead Sea Scrolls

Norma and Edmond Bordeaux Szekely in 1977.

EPILOGUE

This is not the only previously unpublished book of Edmond Bordeaux Szekely to be found after his death; there are a few more which will appear, as soon as I can prepare them, and then his literary storehouse for this century will be complete.

And what a treasury it is! Most readers of this volume know only his writings in English—there are dozens more, in five or six other languages, and an unknown number of early books, long out-of-print. But we shall be content with the eighty-or-so titles available to us at present. If we were able to put into practice the knowledge and practical wisdom contained in even one of his books, we would have advanced our individual evolution immeasurably. And this is why he devoted so much of his abundant energy to writing: not to impress us with his genius (which happened, nevertheless), but to inspire us to live, day by day and hour by hour, the ageless teachings and profound wisdom contained in those hundreds and hundreds of pages which flowed from his brain and heart.

I have dedicated the rest of my life to propagating the books of Edmond Bordeaux Szekely; my fervent hope and desire is that the readers of his books may dedicate themselves with equal ardor to the joyous task of walking the Essene Way of Biogenic Living in their daily lives, and so make eternal the creative flame of his life.

—Norma Nilsson Bordeaux Szekely

CREDO

of the International Biogenic Society

We believe that our most precious possession is Life.

We believe we shall mobilize all the forces of Life against the forces of death.

We believe that mutual understanding leads toward mutual cooperation; that mutual cooperation leads toward Peace; and that Peace is the only way of survival for mankind.

We believe that we shall preserve instead of waste our natural resources, which are the heritage of our children.

We believe that we shall avoid the pollution of our air, water, and soil, the basic preconditions of Life.

We believe that we shall preserve the vegetation of our planet; the humble grass which came fifty million years ago, and the majestic trees which came twenty million years ago, to prepare our planet for mankind.

We believe we shall eat only fresh, natural, pure, whole foods, without chemicals and artificial processing.

We believe we shall live a simple, natural, creative life, absorbing all the sources of energy, harmony, and knowledge, in and around us.

We believe that the improvement of life and mankind on our planet must start with individual efforts, as the whole depends on the atoms composing it.

We believe in the Fatherhood of God, the Motherhood of Nature, and the Brotherhood of Man.

—composed in Paris in 1928 by Romain Rolland and Edmond Bordeaux Szekely

APPLICATION FOR ASSOCIATE MEMBERSHIP
INTERNATIONAL BIOGENIC SOCIETY

Please return to: *I.B.S. Internacional*
P.O. Box 849, Nelson, B.C.
Canada V1L 6A5

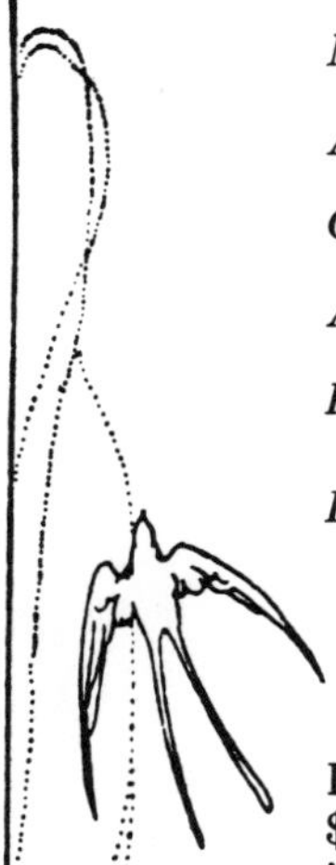

*Name*____

*Address*____

*City, State/Prov., Zip/Code*_____________________________

*Age*____*Profession*__________________________________

*Previous Experience*________________________________

I am interested in:

____*becoming an Associate Member of the I.B.S.*

____*becoming a Teacher of Biogenic Living.*

Enclosed is my annual Associate Membership fee of U.S. $20.00. Please mail me my membership card, your current issue of the Periodical Review, *The Essene Way*, and my copy of *The Essene Way-Biogenic Living*, my "Guidebook," textbook and encyclopedia of ancient wisdom and modern practice. I understand I will receive a 20% discount on all publications as an Associate Member, but only if I order *directly* from I.B.S. Internacional.

Please make your check in U.S. currency out to I.B.S. INTERNACIONAL.

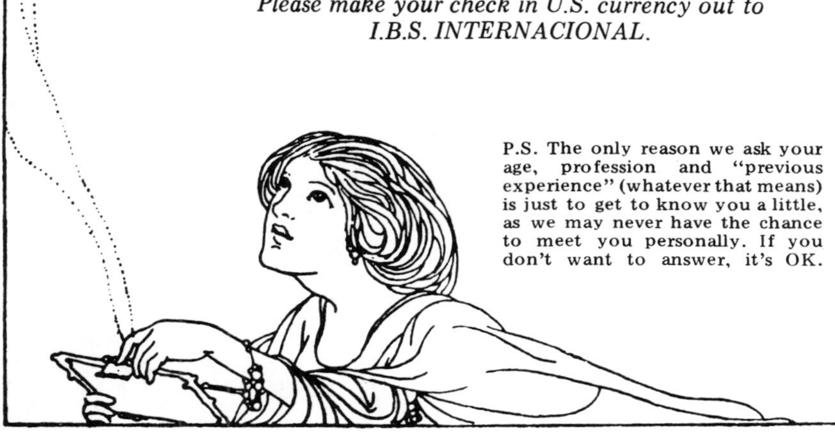

P.S. The only reason we ask your age, profession and "previous experience" (whatever that means) is just to get to know you a little, as we may never have the chance to meet you personally. If you don't want to answer, it's OK.

RECOMMENDED BOOKS FOR STUDY

Many members who have not yet started study groups or live far from book stores and/or like-minded individuals, are interested in a systematic program of home study. The following books are recommended for such a program, and provide an excellent foundation for study of *The Essene Way of Biogenic Living*, especially when coordinated with the methods outlined in *The Art of Study: the Sorbonne Method.*

Please send me the following books:

____*The Essene Gospel of Peace, Book One $1.00*
____*The Essene Gospel of Peace, Book Two $7.50*
____*The Essene Gospel of Peace, Book Three $7.50*
____*The Essene Gospel of Peace, Book Four $5.95*
____*The Essene Way-Biogenic Living $8.80 (free with membership)*
____*The Biogenic Revolution $9.50*
____*The Chemistry of Youth $7.50*
____*From Enoch to the Dead Sea Scrolls $5.95*
____*The Essene Book of Creation $4.50*
____*The Essene Code of Life $3.50*
____*The Essene Science of Life $3.50*
____*Essene Communions with the Infinite $4.50*
____*Discovery of the Essene Gospel of Peace $5.95*
____*The Essene Book of Asha $7.50*
____*Books, Our Eternal Companions $3.50*
____*Cosmos, Man and Society $6.80*
____*Search for the Ageless, Volume One $7.80*
____*Search for the Ageless, Volume Two $8.80*
____*The Ecological Health Garden, the Book of Survival $5.95*
____*The Art of Study: the Sorbonne Method $3.50*

______*Total*
______*Add 15% for Postage & Handling ($2.00 minimum)*

*U.S. $*______________*Total Amount Enclosed*

Please address all orders and correspondence to:

I.B.S. INTERNACIONAL
Box 849, Nelson, B.C., Canada V1L 6A5

*Name*____

*Address*____

*City, State/Province, Zip/Code*____________________________

All orders must be pre-paid. Minimum order: $10.00, minimum postage: $2.00 (foreign orders may be more). Please make check, bank draft or money order in U.S. FUNDS, drawn on a U.S. BANK, with the proper "microencoding" at the bottom of the check, out to I B.S. INTERNACIONAL. Allow 4 weeks for processing, more for foreign orders. Members of the I.B.S. may deduct 20% and *active* Teachers of Biogenic Living may deduct 40% (both must have valid memberships for the current year). *All sales are considered final; please do not return any books.* (If there are any questions, PLEASE WRITE before acting, and please print! Dealers and distributors, write for discount information. All books have always been sent from our warehouse in the U.S., hence the necessity for payment in U.S. funds. Thank you! We send you peace and hope to hear from you soon!